NORTH CAROLINA
STATE BOARD OF COMMUNITY COLLEGES
LIBRARIES
SAMPSON TECHNICAL COLLEGE

AUTHOR'S BOOKS:

1000 Beautiful House Plants and How to Grow Them
Your Homemade Greenhouse and How to Build It
The Old-Fashioned Cutting Garden
Plant Language
1000 Beautiful Garden Plants and How to Grow Them
Your Window Greenhouse
The 5x5 Food Garden

The Pineapple-Top Grower's Handbook

Jack Kramer

Prentice-Hall, Inc., Englewood Cliffs, New Jersey

Book design by Joan Ann Jacobus
Art director: Hal Siegel

The Pineapple-Top Grower's Handbook
by Jack Kramer
Copyright © 1979 by Jack Kramer

All rights reserved. No part of this book may be reproduced in any form or by any means, except for the inclusion of brief quotations in a review, without permission in writing from the publisher.
Printed in the United States of America
Prentice-Hall International, Inc., London/Prentice-Hall of Australia, Pty. Ltd., Sydney/Prentice-Hall of Canada, Ltd., Toronto/Prentice-Hall of India Private Ltd., New Delhi/Prentice-Hall of Japan, Inc., Tokyo/Prentice-Hall of Southeast Asia Pte. Ltd., Singapore/Whitehall Books Limited, Wellington, New Zealand

10 9 8 7 6 5 4 3 2 1

Library of Congress Cataloging in Publication Data

Kramer, Jack
   The pineapple-top grower's handbook.
    1. Pineapple.   2. House plants.   I. Title.
SB375.K7   635.9'34'22   79—12431
ISBN 0-13-676288-3

Contents

1. The Pineapple Plant	9
What Are Bromeliads?	10
Varieties of Pineapples	10
The Pineapple at Home: Grow Your Own	11
How to Pick a Pineapple to Eat and Grow	14
2. Starting Your Own Pineapple Plants	16
Containers for Growing the Pineapple	20
Starting the Pineapple	20
Method 1: In Water	21
Method 2: In Starting Mediums	26
Method 3: In Soil	27
Bottom Heat	27
What Can Go Wrong?	28

3. Keeping the Pineapple Growing — 30
- Soil — 30
- Putting the Plant in a Pot — 32
- Temperature/Humidity — 33
- Watering — 34
- Light — 35
- Feeding — 36
- Insect Protection — 37
- Ventilation — 37

4. Growing and Using the Mature Pineapple — 39
- Care of the Mature Plant — 40
- Flowering — 41
- After Blooming — 42
- Repotting — 42
- Propagation — 47
- Summer Vacation — 53
- Growing Pineapples on Cork or Bark Slabs — 54
- Pineapples in a Greenhouse Situation — 55
- Grooming — 56
- Decorating with Pineapples — 57

5. Bromeliads to Grow with Pineapples — 59
   Ananas — 60
   Aechmeas — 62
   Neoregelias — 63
   Billbergias — 63
   Guzmanias — 65
   Vrieseas — 66

6. Other Plants to Grow with Pineapples — 68
   Orchids — 68
   Cacti — 70
      Epiphyllopis (Rhipsalidopsis) (Easter cactus) — 71
      Rhipsalis (chain or mistletoe cactus) — 71
      Zygocactus (Thanksgiving or crab cactus) — 72

7. Pineapple Nutrition — 73
   Raw Fruit and Juices — 74
   Cooking with Pineapples — 74
   Cutting a Pineapple — 75

1. The Pineapple Plant

The pineapple is from the group of plants called Ananas within the large family Bromeliaceae. Because of their beautiful leaves and colorful flower bracts, many bromeliads have been popular and handsome house plants in Europe for decades. Now bromeliads are becoming indoor favorites in this country, especially since most of these plants are amenable to homes and apartments (never growing too large) and are easy to grow.

Once the pineapple was called the fruit of the gods. Originally from Brazil and Paraguay, where it was called *na-na* (meaning fragrance), the pineapple reached Europe about 1555, and by 1600 or so the plants were being cultivated in fields in parts of India. The English started the pineapple in hothouses in the 1700's, but success was minimal.

For decades the pineapple was considered a delicacy, a delicious fruit for flavor and fragrance, a fruit that enjoyed wide popularity with the rich. Today the pineapple is a valuable commercial crop, mainly grown in Hawaii and Mexico.

🌿 WHAT ARE BROMELIADS?

Bromeliads are epiphytic plants; that is, they grow on trees in the air. All bromeliads like moist, fresh moving air; they do not thrive at all in a stagnant situation. Often bromeliads grow side by side with orchids. The plants have a shallow root system and prefer perching on slabs of wood or branches but will also succeed in pots in home gardens.

Varieties of Pineapples

The groups within the bromeliad family include Aechmea, Ananas, Billbergia, Neoregelia, Guzmania and Vriesea. Such plants as the living vase or urn plant (*Aechmea fasciata*), queen's tears (*Billbergia nutans*), and the flaming sword (*Vriesea carinata*) are now popular house plants.

The bromeliad group known as Ananas contains the commercial

pineapple, scientifically called A. comosus. This is the popular eating pineapple, which is grown commercially in Hawaii and Mexico. This rich fruit is available in markets almost all year round.

There are many variations of A. comosus, noticeable mainly by shape (some are elongated spheres whereas others are round) and outside color (deep green to golden green).

The fruit of the pineapple develops from a dense cluster of tiny lavender flowers that are borne on a short stalk from the center of a rosette of leaves. The flowers combine with the bracts and become fleshy, eventually forming the fruit. The fruit takes only six months to mature and ripen.

The pineapple is greatly resistant to drought; the leaves can accumulate the water from the atmosphere which slides down the leaves into the natural base shape of the plant ready for use in dry spells. The plant is easy to grow and has a short growing season from plant to fruit (only two years), thus making it a good commercial crop.

Although the pineapple can tolerate drought if necessary, it cannot endure temperatures lower than 50 degrees Fahrenheit. It revels in warmth and sunshine.

🌿 THE PINEAPPLE AT HOME: GROW YOUR OWN

You can buy a pineapple plant, but it will be expensive. Actually, there is little need to buy the plant because you can grow your own

LIFE SPAN OF PINEAPPLE/6 Months

9-16 Months

from a pineapple bought in the fruit section of your market—it takes about five minutes to do, as we explain in the next chapter. So instead of throwing away the pineapple top, let it make a plant for you.

Growing a pineapple plant indoors is no more difficult (and perhaps easier) than growing philodendrons. Pineapple plants adapt easily to average home conditions and need only sun and even moisture. Plants should be potted in tightly packed small-grade fir bark, or in a soil-and-bark mix, and the rosette (vase, where leaves join fruit) of the plant should have water in it at all times.

You can also grow the pineapple plant on a piece of wood or cork slab (sold at orchid suppliers).

HOW TO PICK A PINEAPPLE TO EAT AND GROW

Years ago pineapples were available to markets only seasonally—for a few months of the year. Today, you can buy the fruit almost all year round. Most of the pineapples sold have plain green leaves, but occasionally you may find one with variegated leaves or leaves edged red; try to get these because they make such pretty plants—after, of course, you have eaten the fruit.

Because an unripe pineapple will not take root and grow, it is important that you know how to pick a pineapple for ripeness so you can both eat it and enjoy it as a plant. Some authorities say to pluck a leaf from the top of the plant: if it comes out easily, the pineapple is ripe. Other people like to snap their fingers against the side of the fruit. If there is a hollow sound, it is considered ripe. I have my own method for securing a ripe pineapple: I feel and smell. This is hardly esoteric and perhaps somewhat unorthodox. I gently press the bottom of the pineapple with my fingers—if there is some give to it, I am reasonably sure I am getting a ripe pineapple—and at the same time I also inhale deeply. A good ripe pineapple has a heady fragrance. Try my scent and feel method; it works for me and might work for you.

2. Starting Your Own Pineapple Plants

There are three ways of starting your pineapple plant. No matter which method you use, you need little equipment other than a container for your plant, and starting medium or water. The very first thing you have to do is cut off the top of the pineapple about 2 inches below the leaves. Cut with a clean sharp knife in one slicing motion: no ragged edges should be on the pineapple surface, and the surface should be level and flat. Remove some but not all of the leaves. Let the pineapple top dry out and scar over for a few days before planting it.

CONTAINERS/Pickle jar

Clay pot

Decorative tin container

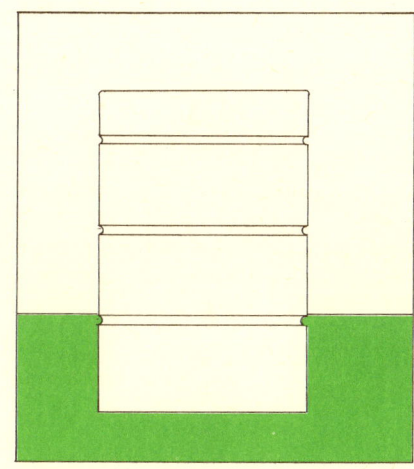

2 lb. coffee can

PLANTING A PINEAPPLE/1. Cut off top of fruit

2. Let dry a few days

🌿 CONTAINERS FOR GROWING THE PINEAPPLE

You can grow your pineapple in almost any type of watertight container or a glass pie plate that is at least 3 inches deep. Glass jars and terra-cotta pots are also satisfactory. Be sure the container is clean and sterile; wash and dry the container thoroughly before planting in it. If you use an old clay pot, be sure to scald and scrub it first so it is thoroughly clean.

Generally, it is best to start the pineapple in a shallow dish or container; a deep pot filled with starting medium or soil can become waterlogged because of too much water in the medium and not enough for the plant. This results in rot that kills the plant.

🌿 STARTING THE PINEAPPLE

Once the pineapple top has dried out and has scarred over (in about 48 hours), you can start it in (1) plain water, (2) a starting medium such as vermiculite, or (3) a porous soil in a clay pot. Let's look at each method in detail.

Method 1: In Water

Insert four small sticks or long toothpicks (shish-kebab wooden skewers work fine) equidistant around the perimeter of the dried pineapple top. Prop the top into a water-filled, large-mouthed glass jar or

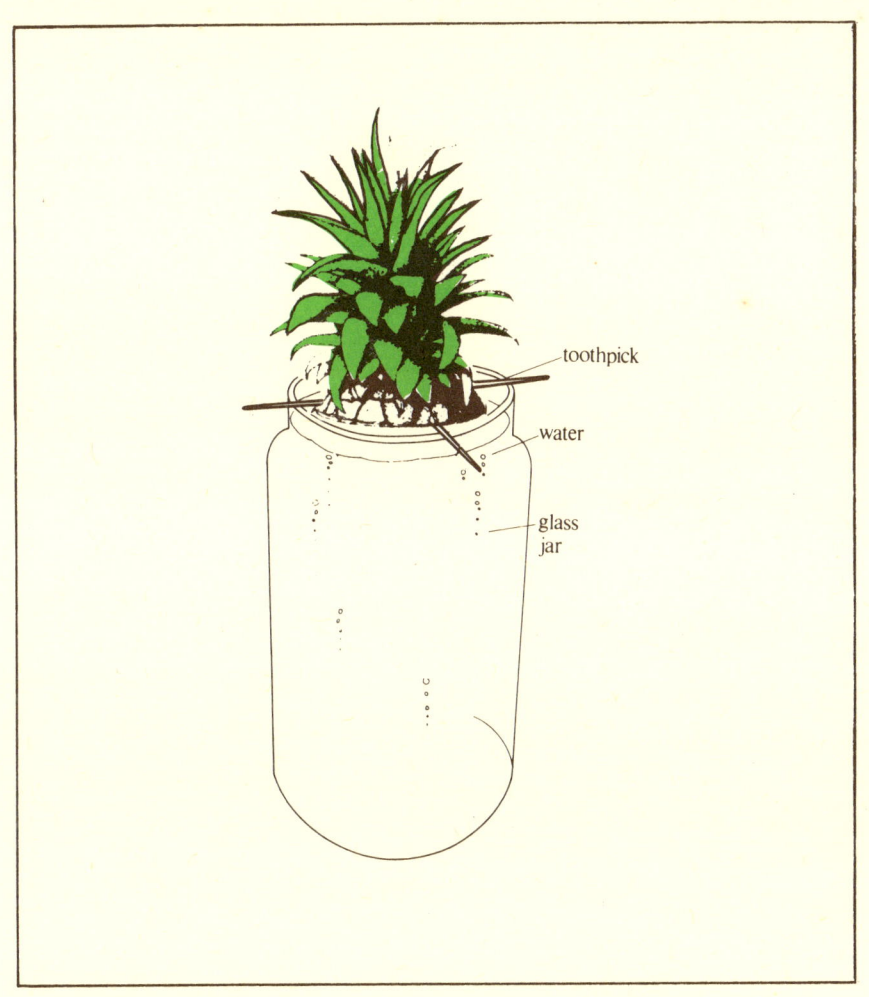

TOOTHPICK METHOD OF GROWING PINEAPPLE
1. Place top in water

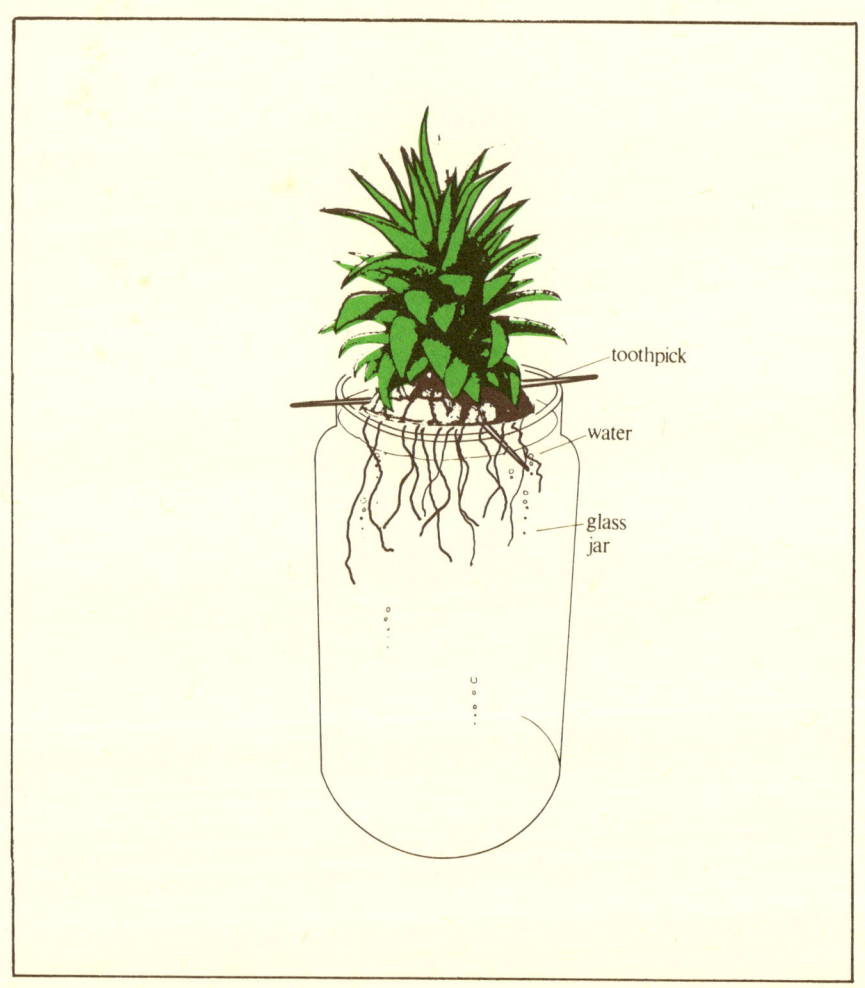

2. Root showing

3. After rooting, replant in soil

THE STARTING MEDIUM METHOD/1. Plant top in starting medium

2. When new leaves appear, pot in soil mixture

similar container so that the bottom 1 inch (or less) of the pineapple is submerged in the water. As the water evaporates over a few days' time, replace it, and keep the fledgling plant in a bright but not sunny window where the temperature is about 75 degrees. When tiny roots start to form, pot the pineapple in soil, as described in the next chapter.

This method is easy and fast, requires no exotic equipment, and is cost-free. The disadvantage is that it is not foolproof: occasionally rot develops at the base of the pineapple, and all is lost.

Method 2: In Starting Mediums

A more acceptable way of starting a pineapple almost guarantees results. Buy some vermiculite or a similar starting medium at the nursery. (These mediums are packaged under various trade names.) Use an aluminum pie plate or a ¾-inch clay pot—any shallow container. The container should have adequate drainage holes. Fill the container with the starting medium to within ½ inch of the top. Now imbed the pineapple 1 inch into the medium, and place the container in a bright but not sunny window.

Keep the starting mix just barely moist, never soggy, or rot will develop. Be sure there is good circulation of air. If humidity is very low, put a plastic bag loosely over the top of the pineapple. Remove the bag when new green leaves appear and put the pineapple in a pot of soil.

This method ensures that a better root system is developed. The disadvantages are that there is some cost involved (the growing medium) and some time is necessary to make the setup.

Method 3: In Soil

This method sometimes works and other times does not, but it is so simple that it is worth trying. Fill a shallow pot—one that has ample drainage—with porous soil and imbed the pineapple top 1 inch into the soil. Keep the fledgling plant in a bright airy place and the soil just barely moist. Watch for rot at the base; if you see any grayish mold, stop watering and let the plant dry out.

This method eliminates the transplanting stage; you simply continue to grow the plant in the same pot. The disadvantage is that sometimes a soil-started pineapple rots at the base and you do not discover this until it is too late.

☘ BOTTOM HEAT

No matter how you started your pineapple top—in water, mix, or soil—it is essential that the plant receive bottom heat if it is to grow properly. Bottom heat is heat at the root or under the plant. I keep my tops in containers on a refrigerator, where there is some heat from the appliance.

Heating cables can furnish a constant bottom heat of about 80 degrees, which is what the pineapple wants. The cables are inserted into the bottom of a conventional container such as a clay pot. These cables are inexpensive and are sold at hardware stores.

If you prefer, simply keep your pineapple top in a location where it is warm all day and all night. Some room or area in your house is probably warmer than other areas, and this is the place to situate your pineapple. Fluctuating temperatures, from day to night, hinder the plant's development when the plant is still small.

🌿 WHAT CAN GO WRONG

No matter which method you use to start your pineapple plant, your attempts at establishing an indoor plantation might go wrong, but don't despair. Basically, here are the things to watch for so your pineapple gets off to a good start and you will feel like a successful indoor grower:

1. In the water method, rot may start because there is constant moisture at the bottom of the cut pineapple. If you see any signs of mushy or grayish growth, rot has started and it is best to begin again. To avoid rot in the first place, keep the waterline just touching the base of the cut pineapple. And to avoid a stagnant odor, add some charcoal chips to the water.

2. In method 2, if you keep the starting medium too moist,

fungus or mildew (sometimes called damping-off) may develop. You can use a fungicide, but the best course of action is just to start over. Also, excessive high temperatures (over 80 degrees) or fluctuating temperatures can cause failure in this method.

3. If you start a pineapple top in soil, be sure it is sterilized soil or again you are opening the door to bacterial or fungus infection. As a precaution when starting a pineapple in soil, place a thin layer of gravel on the top of the sterile soil (soil bought in packages). The gravel will help evaporation of excess water and prevent any rot starting at the base of the plant.

4. In all methods of starting a pineapple, use clean containers and clean knives. It is tempting to reuse a pot, but if you do so scald and scrub the container first. It is tempting to use old soil you happen to have left over from another plant, but if you do there may be some bacteria in it that will thwart the start of the fruit.

5. Even with the best culture and starting methods some pineapples simply will not respond. Don't give up. Start another.

3. Keeping the Pineapple Growing

You have started your pineapple top; now you must keep it growing so it can develop into a stately plant. After a few weeks new green leaves should appear, and it will be time to replant the pineapple. When repotting, you have to consider what kind of soil mix to use and how to repot. Then you must give the plant routine care so it matures properly. You must control the temperature, light, and watering, among other things.

 SOIL

Pot your pineapple in a loose, porous soil, that is, a mealy soil with air passages. You can use standard house plant soil, but add some sand to it—a handful of sand to an 8-inch pot works fine. If you want a special soil for the bromeliad, this is the mix I have used for years with splendid results:

⅓ part packaged soil
⅓ part sand
⅓ part small fir bark

You can also use all fir bark as a potting medium; fir bark is available at suppliers. This is a wood material sold in coarse, medium, or small pieces. It is an excellent medium for epiphytic plants such as bromeliads and orchids. The potting procedure is more or less the same as for soil; the plant is centered in the container and the fir bark is packed around the plant to support it upright.

The advantages of fir bark potting are that air reaches the roots of the plant and of course there is no mess as there sometimes can be with soil. Also, it is difficult to overwater a pineapple in bark: the material dries quickly.

There are so many packaged soils that it makes good sense to read the packages to determine just what is in the sack. Some soils do not have many nutrients. A good soil is porous and mealy to the touch and smells of the woods (humusy). Because packaged soil is just that —packaged—you cannot smell it, but you can feel it. Squeeze the sack: it should have some give to it.

🌿 PUTTING THE PLANT IN A POT

When your plant is ready for potting, choose a container that is about 8 to 10 inches in diameter. Clay pots are very satisfactory and are better for pineapples than, say, plastic pots. Plastic is lightweight, so often the weight of the plant topples the plastic container. No matter what kind of container you use, be sure it has drainage holes at the bottom so excess water can escape. The pineapple plant is especially fussy about water: it likes its roots moist but never soggy.

Prepare the container with small pot shards (pieces of clay pots) at the bottom. A 1-inch bed is fine. Or use a gravel bed. Either shards or gravel will ensure good drainage. Now add a mound of soil mix and center the plant. If it is too low in the pot, add more soil. If the plant is too high, take out some soil. Allow a 1- to 2-inch space at the top of the container for watering. Fill in and around with soil; keep tapping the bottom of the pot against a tabletop to settle the soil mix. Do not bury the crown in soil, or rot may occur. Firm the soil with a blunt-nosed stick or with your thumbs, pressing down the soil. You want it to be firm but not tightly packed.

If you are using bark for potting the plant, follow the same procedure but remember that bark has very little nutrients in it and the pineapple will have to be fed with plant foods more frequently—about every other watering—than if it were potted in soil.

Water the new plant thoroughly; let the water drain, and then water the plant again. Now the pineapple is ready for a new location.

🌿 TEMPERATURE/HUMIDITY

Your pineapple will grow best at stable, warm temperatures, about 80 degrees. When the plant is a few months old, slight fluctuating difference between day and night temperatures is then desirable. But pineapples should never be grown cool; try to maintain a minimum night temperature of 65 degrees.

For best results maintain a humidity of 40 to 60 percent for the plant. Plants subjected to a lower humidity in the home do not do well; they grow slowly and struggle to survive. Most homes do not have this much moisture in the air unless there is an automatic humidifier as part of the heating system. If you have many plants growing together, the plants themselves will furnish humidity because plants give off moisture through their leaves.

However, if you have only a few plants and your home does not have an automatic humidifier, the best way to provide humidity is to spray plants with a mist of water. Use a spray bottle (sold at supermarkets) once or twice a day to mist the leaves of the plant. If you are not home enough to do this, then put containers on a bed of gravel and keep the gravel bed moist but never soggy. Use a plastic tray or deep pie plate to hold the gravel. The evaporation of water on the gravel furnishes the necessary humidity for the plant.

🌿 WATERING

Watering plants is just a matter of common sense—it is not a question of when you water but how much water you give plants. If the container's drainage is good (and it should be), you can water pineapples heavily: let water run out through the pot and accumulate in the saucer. Forget the old wives' tale that water in the saucer will kill a plant; it does not.

However, if drainage is not good, then you must water judiciously. Otherwise soil will become waterlogged and eventually roots will rot and die and the plant will suffer. When you water the plant, *really* water it, and then let it dry out somewhat before watering it again. Usually this means watering every other day or at least every third day in spring and summer and twice a week in the fall and winter. Although it is wise to water somewhat less in winter, remember that artificial heat dries out plants considerably, so if you keep your home very warm, then you must adjust the watering schedule for your pineapple plants.

The size of the pot affects how fast the soil dries out: large pots hold water longer than small pots. So once again some common sense is necessary for giving plants the right amounts of water. Incidentally, icy cold water shocks plants, so it is much better to use tepid water. To make water tepid, let it stand in a watering vessel overnight; this also helps evaporate any chlorine in the water. (Some large cities use a great deal of chlorine in their water supplies, and chlorine may affect plants adversely.) With pineapples you must also water the center of the plant. This rosette holds water so the plant can take care of itself if you forget to water it.

🌿 LIGHT

Few plants grow without light; pineapples like a bright but not direct sunny location. Place plants at south or west windows where they can get good bright light and a few hours of sun. You can tell whether your plant is getting enough light by the color of the leaves: they should be shiny dark green, not pale.

If light is a problem in your home, consider growing the pineapple top under artificial light. There are many plant-growth fluorescent tubes on the market, sold under various trade names. You will need two 40-watt lamps. Or consider one of the newer plant lamps that fit into regular lamp sockets; this light is aimed at the plant. These work very well for one or two plants and furnish the necessary red and blue rays plants need for good growth. Lights should be on 14 to 16 hours a day during the spring and summer, about 12 hours during the fall and winter.

If you have no plant-growth lamps and do not want to buy any, at least use a standard 60-watt house light. Be sure to keep enough distance from the lamp to the plant so heat does not desiccate the plant. The safe distance is about 30 inches.

🌿 FEEDING

Feed your pineapple plant every other watering during the spring and summer, using a 10-10-5 plant food, which is neither too strong nor too weak. In fall and winter, feed only once every six weeks. The easiest way to feed the plant is to sprinkle a little granulated plant food on the soil and then apply water. Liquid fertilizers are also satisfactory, but never use strong plant foods because they can cause more harm than good.

Foliar feeding—watering and feeding plants by spraying the leaves—is also acceptable, but I do not stress it too much because it is a bother (you need special plant food and apparatus to do the job), especially indoors, where you may end up spraying half your kitchen.

🌿 INSECT PROTECTION

In all my years of growing bromeliads I have never seen insects on them. The leaves are tough and insects usually migrate to more easy fare. However, if insects such as mealybugs or aphids do attack your plants, do not panic. There are household remedies that will get rid of them without your having to resort to poisons.

Here are the three household remedies I have used:

1. Make a solution of one-half bar of laundry soap and 2 quarts of water; mix thoroughly, and douse plants with it. Then rinse the plants with warm water.

2. Put rubbing alcohol on cotton swabs and swab the insects. Repeat every fifth day.

3. Use a solution of old cigarette tobacco and water that has steeped for a few days.

If you must resort to other preventatives, use only rotenone or pyrethrum, because both are natural plant repellents.

🌿 VENTILATION

We all need air, and plants are no exception; they like good ventilation, or in more concise terms, a good circulation of fresh air. It is important that the air in the growing area never become stagnant—remember that pineapples are epiphytes (plants that thrive on air). In spring and summer there is little problem in providing a fresh free flow of air—windows will probably be open—but it is just as essential to provide good air circulation in fall and winter. Keep a window slightly open in the growing area if at all possible. Or if impossible, at least have a small fan going at low speed to keep air moving. (Small 6-inch fans are sold by suppliers.)

In spring and summer you might shut your house up tight and flip on the air conditioner, with no great concern whether your plant will tolerate this artificial coolness. It will but it won't like it for long periods. The essential factor of good growth for epiphytic plants is again a free flow of fresh air.

I am not asking you to keep your air conditioner off. We humans need our comfort too. But at least temper it a bit by opening a window occasionally.

4. Growing and Using the Mature Pineapple

Once you have your pineapple plant growing, you want to keep it healthy and handsome. This requires routine care, but different care than what was needed to bring the plant through the first few months. Now it is a question of keeping it healthy, not just growing. In Chapter 3 we potted our fledgling pineapple plant; here we describe how to raise the mature plant and also discuss how it may be used to accent the home.

Most plants have three phases of care: the initial starting, the adaptation to the new environment, and the maturity. The pineapple's life span is about two years, and during this time it can be a handsome rewarding plant provided it is in good health.

🌿 CARE OF THE MATURE PLANT

It might take a few months to find out just where the pineapple grows best in your home. A few feet one way or another can mean the difference between a handsome plant that thrives and one that just sits. Why? The air circulation may be better in one area than another, the temperature between daytime and nighttime may be less severe, and light may be better.

For optimum growth keep the pineapple plant in a bright and sunny place; at least a few hours of sun a day will keep leaves colorful and healthy. An east or south exposure is fine. Be sure the air circulation is good; there should be no hot air blasts or air-conditioning drafts to strike the plant—both these elements can harm and even kill a plant.

During warm weather you should mist the plant daily. This adds moisture to the air, something plants like, and keeps the leaves free of dust and soot. Once a month wash the leaves with a damp cloth to further ensure that they are clean and shiny. And be sure to check that leaves are free of any insects or larvae.

Once every three months take your pineapple plant to the kitchen sink and submerge it to the rim of the pot in water; this helps leach the soil of any toxic salt buildup from feeding and brings any hidden insects to the surface of the soil where you can see them. Leave the pot in the sink for about 30 minutes or until you see air bubbles form on the soil.

For maximum growth, feed your plant every other watering with 10-10-5 plant food from April until October. Feed only occasionally the rest of the year, but do keep the soil evenly moist.

Every month or so trim away dead leaves; the pineapple naturally loses its leaves as it gets older. Cut dead leaves with a sharp knife as close to the crown of the plant as possible.

🌿 FLOWERING

Your pineapple plant will probably flower on its own if it has had good care, that is, ample light and water. You will see a red bud; this bud eventually opens to show hundreds of tiny blue velvet flowers in a tufted crown. The flowers last only a day or so. Each remaining flower bract develops into one segment of the fruit.

Pineapples indoors seldom form fruit on their own without some help. The help is in the form of an apple. The apple releases ethylene gas, which forces the pineapple to flower and fruit. To use the apple method, place a plastic bag over the plant and set a ripe apple inside the bag; now tie or tape the bag closed.

Remove the apple and bag in four or five days; in a short time new leaves will start forming from the center of the pineapple plant.

Then rows of pineapple fruit will appear on the bottom of new leaves. The fruit will be golden yellow and about 5 inches above the old plant on a robust stalk. Stake the plant now with sticks so it does not topple over If no fruit appears, repeat the apple treatment.

🌿 AFTER BLOOMING

Once the flower and fruit appear, the pineapple plant dies, but not before nature provides little babies called offshoots or *kikis*. When these offshoots are a few inches high, they can be severed from the mother plant with a sharp, clean knife and potted in sand or starter mix, as prescribed for the top of pineapples in Chapter 2. Thus you will have more free plants from nature.

🌿 REPOTTING

During its life with you the pineapple will require two pottings after it has been growing. The first potting is done when new leaves appear, about a month or so after you have started the plant. This was described in Chapter 3. The second potting, or repotting, is done about nine months to one year later. Here is the way to repot a pineapple plant.

REPOTTING PINEAPPLE/1. Add soil to the pot

2. Remove from soil mixture

3. Insert into soil

4. Fill in with soil and water

Put on gloves. Remove the plant from its old container by knocking the edge of the pot against a table to loosen the rootball. Now remove the plant from its pot by teasing it loose; juggle it around to do this. Do not *pull* the plant out of the old pot or you will hurt the plant. If you cannot manage to get the pineapple from its old container, break the pot with a hammer blow.

Once you have the plant out of the old housing, crumble away dead, loose soil and trim away any root tips that are brown (*not* the white ones). Now take a 5- or 6-inch pot, put shards in the bottom, and place a mound of soil in the pot. Then center the plant and add soil as you did for the first potting. Water the repotted plant thoroughly, and then water again. The pineapple is repotted!

If you have problems removing the plant because of the spines (even with gloves a spine can tear through the material), put the plant on its side, pot and all, and wrap several sheets of newspaper around the plant. Then grasp the collar as described and tease the plant from its container.

🌿 PROPAGATION

The word propagation sometimes scares the beginning gardener but it means simply getting more plants from a single plant. Nature is so free with her bounty that there are numerous ways to multiply your plants.

PROPAGATION/1. *Cut off-shoot from mother plant*

2. Insert in soil mixture

3. Water

4. After 5-6 weeks, repot

Propagation is by vegetative means rather than by seed. The pineapple produces suckers, slips, or crowns. *Suckers* are shoots, sometimes called branches, that are borne either from the axils of the leaves or from the stem of the plant near the soil. *Slips* are shoots that grow from the fruit stem, and *crowns* are the rosettes of short leaves which grow on top of the fruit.

Generally, suckers are the preferred method of getting new pineapple plants from old ones (occasionally, suckers are also referred to as offshoots). Crowns and slips take much longer to bear fruit than do suckers.

When the offshoot or sucker is an inch or two long, gently twist it from the mother plant. It should come off easily; if it does not, then cut it with a sharp sterile knife. Pot the tiny plant in a shallow container of starting medium as described in Chapter 2. Do not bury the base of the plant too deeply. Transplanting will be necessary when the pineapple is about 4 to 6 inches tall; place in containers of fir bark or in a soil-sand-fir bark mix.

Occasionally, I have allowed the offshoot to remain on the plant; eventually, as the mother plant wanes and dies, the offshoot takes over and grows to a fairly large size. Then the mother plant can be severed and removed. This is not the ideal way of growing the offshoot but it will work if you are forgetful.

If you use slips or crowns, allow them to dry off for a few days before planting them. Crowns can be rooted in a starting mix or in water if there is good bottom heat, and they may be started in soil as well. All three methods of starting the crown are described in Chapter 2.

🌿 SUMMER VACATION

If you have a place for your pineapple outdoors in summer, by all means move it to a porch or patio or even a windowsill. Fresh air and rain will do a great deal to keep the plant in tip-top shape; in fact, the pineapple thrives outdoors. In most parts of the country you can put the plant out in mid-May and return it to the house after Labor Day. Remember, the minimum temperature for the plant should be between 50 and 55 degrees Fahrenheit. If you forget to take in the plant, one or two cold nights can harm it greatly.

Now if you are going on vacation, what do you do? If you are leaving for a week or ten days, don't fret. See that the plant has some air (keep a window slightly open if possible) and water it thoroughly before you leave. Soak it and soak it again. Generally, the pineapple can survive your ten-day hiatus without injury.

If you are to be gone longer than two weeks, make arrangements to have someone come in and water your plant (you will probably have more than one plant anyway) or at least take it to a friend's apartment and ask him or her to plant-sit.

🌱 GROWING PINEAPPLES ON CORK OR BARK SLABS

If you don't want your pineapple in a pot, consider growing it on a cork or bark slab, which you can get at suppliers. The advantage of slab growing is that, since roots are not completely covered, the plant gets air from all sides and thus grows better. Generally, the slab is mounted on a wall or suspended with thread or chain from a ceiling. There are several hanging devices and methods of placing the slab on a wall.

Though the slab method of growing the plant is a good one, it is not that easy to accomplish. For example, first you must make a tiny bed of osmunda fiber and attach it to the slab. (Osmunda fiber is available from mail order suppliers.) Wire the osmunda to the slab and then wire the base of the plant to the osmunda—sort of a piggyback affair. Use coated plastic wire or string; conventional wire can injure the plant crown. Once you have accomplished this wire-on-wire method and the pineapple plant is attached securely (and it must be secure or else all is lost), let it grow this way for several months. After that you can remove the wire or string because the plant has rooted into the bark or cork.

With slab growing, care is somewhat more demanding than if the plant were in a pot. Frequent watering—every day—is necessary, especially during warm months, and of course excess water will drip on floors. You can put a drip tray on the floor underneath the plant—a

Growing and Using the Mature Pineapple   55

pie plate or some container—but that is hardly esthetic. The ultimate solution is to have a floor that is impervious to water, such as tile or vinyl, but this is not always possible. So if you have a conventional wooden floor or carpets, the only alternative is to water the plant at the sink—hardly a pleasure and usually a chore. Still, it works and you can do it. It just depends how much you like your plants and how much time you have.

🌿 PINEAPPLES IN A GREENHOUSE SITUATION

Growing plants in greenhouses—conventional or window types—is popular and pineapples make good under-glass subjects. Originally, they were considered stove or greenhouse plants in England, and in a controlled environment you *can have* a small plantation.

Be sure plants in a greenhouse situation get adequate ventilation; many times greenhouses are too moist, which can be detrimental to plants (excessive humidity causes fungus diseases). While the pineapple can tolerate some coolness at night, it should never be grown lower than, say, 65 degrees. If your greenhouse has low temperatures (and some do), it is best to grow the plant in the home.

Keep pineapples in the sunniest part of the greenhouse; they tolerate and, in fact, benefit from a great deal of sunlight. Generally, they do not require the shading that other greenhouse plants need. However, watch the leaves to be sure they do not get scorched. Conditions vary in each greenhouse in every part of the country.

🌿 GROOMING

After a time, all plants need some grooming—trimming, leaf washing—and the pineapple is no exception. With many pineapples it is a natural tendency for lower leaves to turn brown after a time, and this is no cause for alarm. Put on your gloves, take a pair of trimming shears, and remove the lower leaves. After cutting, dust the open wounds with powdered charcoal to seal the cuts so bacteria cannot get to the plant.

Every few weeks give your plant a quarter turn: this helps keep the plant growing uniformly in rosette shape rather than becoming lopsided through remaining in the same position all the time. The quarter turn takes but a second and is worth the time to have a handsome plant.

Pineapples benefit greatly from a misting of water on the leaves; this keeps off dirt and soot and keeps the leaf pores open to absorb moisture. In addition to misting every so often, take a damp cloth and wipe each leaf to make it shiny and handsome. Do not use leaf-shining preparations—they invariably clog leaf pores and thus may hinder growth.

🌿 DECORATING WITH PINEAPPLES

The well-grown pineapple has a beautiful rosette shape and makes a fine and colorful indoor accent. The pineapple can decorate any room, but I have found that they look especially attractive in kitchens and bathrooms. In baths they add just the right note of greenery, and they grow well there because of the ample humidity and heat. You can also put two or three pots at a living room window for a nice display, or set a plant at the corner of a desk. To dress up the pot, slip it into a handsome cache pot or any other decorative container. Plants on shelves are very attractive and make nice decorative fillings, or you might want to use hanging containers for your pineapple.

If you want an exceptional display for a room, group three pineapples in a large tub (about 20 inches in diameter), and grow them in this manner. This creates a forest of leaves and looks especially attractive when viewed from above. If you put the container on a low plant stand in the corner of a living room, for example, I am sure you will receive some comments. Three plants grown together always look better than a single plant.

Pineapples grown with succulent plants such as echeverias, agaves (small ones), and aloes also can make a handsome display in a large shallow dish, somewhat like a dish garden arrangement. Place a few well-selected stones in the dish garden and sprinkle some sand or gravel on the top to create an attractive plant grouping.

The pineapple looks even better when used with well-suited companions. Other bromeliads, such as Guzmania and Aechmea, complement the ruggedness of the pineapple and take the same type of care. Orchids and several cactus also look good with bromeliads. The next two chapters discuss companionable bromeliads and other playmates for your pineapple.

5. Bromeliads to Grow With Pineapples

Once you have started your own pineapple plant or a few, you might want to try some other members of this family. In the Ananas group there are two plants that are grown as house plants, and both are fine additions to any plant collection. *A. variegatus* has handsome yellow and green leaves tinged with pink. *A. nanus* is known as the dwarf pineapple and never gets larger than 12 to 18 inches.

From other genera in the Bromeliaceae there are popular house plants such as *Aechmea fasciata* (the living vase or urn plant) and Neoregelias, exquisitely colored rosette-type plants that turn brilliant red at the center at bloom time.

Let's look at all these plants and see how to grow them as additions to the pineapples you have grown on your own.

🌿 ANANAS

A. variegatus (rainbow pineapple) is perhaps the most spectacular in the Ananas group. This plant, slightly larger than the commercial pineapple *A. comosus*, is known for its beautiful yellow and green leaves tinged with pink. It requires a large tub to grow well; use a container at least 14 inches in diameter and a growing mixture of one-third sand, one-third soil, and one-third small-grade fir bark. This combination makes an ideal potting mix.

The rainbow pineapple grows quickly in summer and requires a lot of water at this time; spraying the leaves with water also helps growth. It likes a somewhat brighter and sunnier place than *A. comosus*.

A. variegatus matures within two years. At the end of a year or 14 months it should be repotted. This is not as easy a job as you may think, because *A. variegatus* is somewhat thorny: the leaf edges are sharply scalloped.

A. variegatus does not fruit as easily as *A. comosus* and most likely will not bear fruit indoors, although some of my friends claim theirs did. The flower bract is similar to that of the pineapple plant: large crowns of small pink flowers in white bracts. The crown lasts for months.

Aechmea Fasciata

The other available pineapple plant is the miniature called *A. nanus*. This plant, with its plain green leaves, is hardly a show plant, but it does make a satisfactory indoor subject. It bears small inedible pineapples about 2 inches across.

AECHMEAS

These plants, known as the living vase or urn plants, are, as their common name implies, vase shaped. In their native habitat they are mostly tree-growing. *A. fasciata* (silver urn) is perhaps the best known for indoor culture. It is a handsome plant, with frosty green and silver leaves and a pretty flower crown of red and white.

Unlike *A. comosus*, which is somewhat stingy with its offshoots, the living vase plant produces dozens of little babies after it matures in about two years. To grow *A. fasciata*, use a soil mixture of equal parts of soil and fir bark. Select a pot that is somewhat small for the plant itself—a 5- or 6-inch container is fine because *A. fasciata* likes to be crowded in the pot.

Give the silver urn good light and even moisture all year. Generally it is an easy plant to grow indoors and makes a handsome decoration. This is certainly a good choice for the gardener who says, "I can't grow anything."

Another favorite Aechmea is *A. chantinii*, sometimes called queen of the bromeliads. This is a large—up to 40 inches—vase-shaped plant with a spectacular flower bract of red and yellow. Like other plants in this group, it is easy to grow.

🌿 NEOREGELIAS

I have deliberately selected the Neoregelia group as another example of bromeliads because they grow in a somewhat different manner from the pineapple plant or the Aechmea. These plants are flattened rosettes of leaves, and the center turns a dark red at bloom time. Flowers are tiny—hundreds of pink blooms in the center of the plant. These plants are grown for their foliage and are sometimes called fingernail plants or blushing brides. Some Neoregelias also have red leaf tips.

Neoregelias like a bright but not sunny place and should be grown in equal parts of soil and fir bark. Their cultivation is similar to that of the pineapple plants.

🌿 BILLBERGIAS

These decorative plants are mostly large, with multicolored foliage and bizarre flowers. The plants are perfect for indoor growing. They have gray-green, silver-green, or purple leaves. The flowers are small, but the colorful bracts are striking: red, pink, or purple. Pot Billbergia in fir bark or an osmunda and soil mix. Give plants bright

Neoregelia Spectabilis

light, and keep their vases filled with water. You can propagate Billbergias by offshoots.

B. amoena grows to 16 inches and has shiny green leaves. There are rose bracts in the spring or summer.

B. morelii grows to 10 inches and has green leaves. There are blue flowers in red sheaths in summer.

B. nutans (queen's tears) grows to 30 inches and has chartreuse, pink, and cerise flowers in the winter.

B. pyramidalis grows to 24 inches and has golden green leaves. There are orange-pink flowers and bracts in the summer. 'Fantasia' grows to 24 inches. This robust hybrid has multicolored foliage, red bracts, and blue flowers in the fall.

B. zebrina grows to 40 inches and has gray-green foliage flecked with silver. There are cascading stems of rose bracts, usually in the summer.

GUZMANIAS

These striking plants have rosettes of leaves and small flowers hidden in vivid bracts that stay colorful for four months. Give Guzmania bright light. Pot them in osmunda, and keep the osmunda wet but never soggy. The plants like 50 percent humidity. These gems make fine table decoration.

G. berteroniana has a 20-inch rosette of wine-red leaves with yellow flowers in the spring.

G. lingulata has a 26-inch rosette of apple green leaves with star-shaped orange flowers all summer.

G. monostachia has a 26-inch rosette with red, black, and white flowers in the fall.

G. musaica has a 20-inch rosette of dark green and red-brown leaves, with white flowers in the fall.

G. zahnii has a 20-inch rosette of green leaves, with red and white flowers in the summer.

🌿 VRIESEAS

Feathery colorful plumes that last for several months make the Vrieseas ideal north-window plants. Some kinds have pale green leaves; others have dark green foliage marked and banded in brown. All Vrieseas have rosette growth. Pot plants in equal parts of osmunda and soil (Vrieseas are terrestrial), and keep the vase filled with water. Do not fertilize or spray with insecticides.

V. carinata (lobster claws) grows to 18 inches and has smooth pale green leaves and a yellow and crimson sword.

V. hieroglyphica grows to 30 inches and has a green-banded rosette with darker markings and tall yellow flower spikes.

V. malzinei grows to 12 inches and has plain green leaves and a cylindrical orange flower spike.

V. splendens grow to 20 inches and has green foliage with mahogany stripes and orange swords on erect stems. This plant does not produce offshoots, but new plants push up from the center of mature growth.

6. Other Plants to Grow With Pineapples

The conditions we have described for growing pineapples are well suited to such companion plants as orchids, some forest-dwelling cacti, and a few miscellaneous plants that like airy and somewhat moist conditions. In this chapter we look at other plants that you can grow in the same space as the pineapple—that is, a growing area with good air circulation, some sun, and somewhat moist conditions (about 30 to 40 percent humidity).

ORCHIDS

In their natural habitat orchids grow side by side with bromeliads.

Here are fourteen orchids I grow in the same area as my pineapple plants:

Aërides crassifolium, only 10 inches high, has amethyst-purple flowers.

A. multiflorum, a dwarf variety, has many small rose flowers with darker spots on the lip.

A. odoratum, as the name implies, has a powerful but pleasing scent; the fragrance will fill your house. The small flowers are waxy white, blotched with pale magenta.

Cattleya forbesii carries two to five greenish yellow flowers that are 3½ inches across and whose lips are yellow streaked with red on the inside.

C. luteola is a small plant (6 inches tall) with a 2-inch pale yellow flower that has a white lip and sides streaked with purple.

C. skinneri grows to about 30 inches and produces two to eight rose-purple flowers about 3 inches across.

Cycnoches chlorochilon has the largest flower of the genus, measuring 7 inches across. The flowers are yellow-green, almost chartreuse, with a creamy white lip blotched with dark green. Because of the waxy texture of the inflorescence, it looks artificial, but the heavy morning fragrance, clean and spicy, is delightfully real. Free-flowering and most adaptable to window culture, this species usually retains its foliage until after flowering.

Cypripedium insigne has brown-veined apple green flowers of a shiny waxy texture. Many varieties are available in a wide range of color.

Epidendrum aromaticum, about 12 inches tall, produces greenish white flowers powerfully but sweetly scented.

E. Obrienianum has 1-inch flowers clustered at the top. There are many varieties, and colors range from pink to lavender to brick red.

Lycaste aromatica is a dwarf, rarely exceeding 12 inches, and produces four to eight brilliant orange-yellow flowers about 2 inches across with a heavy cinnamon scent.

Oncidium ampliatum has turtle-shaped pseudobulbs and small spray-type flowers of yellow and red-brown. If the same spike is cut below the last node when the flowers fade, the plant sometimes produces a second flower stalk, as it did for me.

O. splendidum, the most commonly grown orchid, has solitary 12-inch long cactus-like leaves and vibrant yellow flowers barred with brown; the lip is yellow, large, and broad.

Trichopilia suavis is about 14 inches tall and produces cream-white, 6-inch flowers that are spotted red. The trumpet-shaped lip is spotted pink and orange; the flower has a pleasing hawthorn scent.

🌿 CACTI

I know that the vision of a cactus will make you think of deserts and sand and dry conditions, but although many cacti do grow in these environments, some, such as the Christmas cactus and various other species, like the epiphytic atmosphere.

Epiphyllopsis (Rhipsalidopsis) (Easter cactus)
These 12- to 20-inch epiphytes bloom when they are very small, with bright red, pink, or orange-pink flowers. Grow them in sun in the fall and winter with a 55 degree temperature and 12 hours of uninterrupted darkness. In the spring and summer, water the plants freely and regularly. The Easter cactus likes 60 percent humidity. Propagate the plants by cuttings.

R. gaertneri is the true Easter cactus. It has bright red flowers.

R. rosea is a small plant with soft pink flowers in the spring. 'Orange Spring Beauty' has a delicate color for early spring.

Rhipsalis (chain or mistletoe cactus)
This fine group of spineless, generally epiphytic, cacti has pendant growth to 36 inches and handsome colorful berries in winter. Pot Rhipsalis in osmunda or fir bark or an osmunda-soil mix, and grow it moist in the summer but somewhat dry in the winter. Give the plants bright light. Propagate plants from cuttings.

R. burchelli (mistletoe cactus) has cream-colored flowers and pink berries.

R. capilliformis has cream-colored flowers and white berries.

R. paradoxa has three-angled stems, white flowers, and red berries.

Zygocactus (Thanksgiving or crab cactus)

These popular plants from Brazil bring color to the fall indoor garden. They have toothed branches and bear exquisite, dainty blossoms from late October into December. In general, these epiphytic 24- to 30-inch plants need sun in the fall and winter and bright light in the spring and summer. Keep soil moderately moist, except in the fall, when roots should be somewhat dry. In the fall grow plants quite cold (55 degrees) with 24 hours of uninterrupted darkness for one month to encourage bud formation. The Thanksgiving cactus likes 50 percent humidity. Pieces of the stems root easily for propagation.

Z. 'Amelia Manda' has large crimson flowers.

Z. 'Gertrude W. Beahm' has bright red blooms.

Z. 'Llewellyn' has orange flowers.

Z. 'Orange Glory' has pale orange flowers with white throats.

Z. 'Symphony' has delicate orange flowers whose petals are white at the base.

7. Pineapple Nutrition

The pineapple is more than a plant you can grow at home and a fruit you buy at supermarkets. It is also a very nutritional food, good for you and good to eat.

Pineapples and pineapple juice are useful providers of vitamins C, A, and B_1. A slice of pineapple is only 50 calories, a glass of juice only 135 calories. The pineapple is especially rich in minerals—potassium, chlorine, sodium, phosphorus, sulphur, calcium, iron, and iodine.

❧ RAW FRUIT AND JUICES

You can eat pineapples out of hand and get a good supply of vitamins and minerals, but if you juice the pineapple you can get more health from it. One medium-size pineapple yields about two cups of juice, and with today's new juice extractors, making your own pineapple juice is a snap. Peel, quarter, and feed the fruit into the machine and you have a refreshing, healthful drink both in summer and in winter.

Pineapple juice blends well with other juices such as apple, grape, peach, and even beets! Indeed, there is a whole new world of taste treat when you combine pineapple juice with other juices. And pineapple juice can be a refreshing treat any time of the day—for breakfast, between meals, and even at bedtime.

Canned or frozen pineapple juice does not contain the high concentrate of vitamins and minerals—more than half are lost in processing—so it is the fresh fruit you want, eaten out of hand or taken as a juice.

❧ COOKING WITH PINEAPPLES

In addition to beverages and desserts, there are dozens of recipes that use pineapple in cooking. Fresh pineapple can be cut into wedges and added to roasts to give this dish new zip; it can be sliced and cooked with ham or used equally well with pork as a side dish. You can have roast chicken with pineapple, and so forth. I have not included specific recipes here: what I do is make my usual roast or chicken dishes and in the last few moments I add fresh sliced pineapple or some pineapple juice.

You can also use fresh pineapple chunks as a garnish for meat dishes; the sweet and tart taste complements almost any food and at the same time adds a decorative note to the serving.

🌿 CUTTING A PINEAPPLE

Some people put a pineapple on its side, slice off a section of the fruit, and then cut away the peel—and this is fine. However, if you are using the fruit as a dessert you might want to serve it as follows:

Split the pineapple vertically in four equal pieces (leaves intact). Using a sharp knife, cut through the pineapple and the leaves. Now place the segments horizontally on a tray and with a paring knife carve a line slightly above the skin and below the meat of the fruit. Then slice vertical chunks. Dust with powdered sugar and serve with toothpicks for a delightful and attractive dessert.

Finally, the pineapple can be used as a container for a fruit punch or an alcoholic one, whatever your preference. In this case the top is cut off, the first 2 inches are sliced off, and the meat of the pineapple is removed. Taking out the fruit is not all that easy but it can be done. I use a sharp paring knife and make a small circular cut in the center of the pineapple first; then I chisel away the remaining fruit until I have the pineapple shell. Your favorite tropical beverage concoction can be put into the shell, just as they often serve pineapples in the islands.

So whether you eat it or grow it, the pineapple is indeed a diversified fruit, one you can enjoy as a food or as a plant for nominal cost.